21st Century Junior Library

MAKING A DIFFERENCE WITH THE WORLD WILDLIFE FUND

Katie Marsico

How Can I Help?

Published in the United States of America by:

Cherry Lake Press
2395 South Huron Parkway, Suite 200, Ann Arbor, MI 48104
www.cherrylakepress.com

Reading Adviser: Marla Conn, Read With Me Now
Photo Credits: © Hung Chung Chih/Shutterstock, cover, title page; © Rene Van Den Berg/Dreamstime.com, 5; © Juergen Freund naturepl.com, 6; © littlenySTOCK/Shutterstock, 9; B. Bartel/USFWS, 10; © Yuri Arcurs/Dreamstime.com, 12; © Rozenskip/Dreamstime.com, 14; © goodluz/Shuttersock, 15; © Joe Sohm/Dreamstime.com, 17; © Duncan Noakes/Dreamstime.com, 18; © Carlos Velázquez/Dreamstime.com, 20; © EQRoy/Shutterstock, 21

Cherry Lake Press is an imprint of Cherry Lake Publishing Group.

Library of Congress Cataloging-in-Publication Data

Names: Marsico, Katie, 1980- author.
Title: Making a difference with the World Wildlife Fund / Katie Marsico.
Description: Ann Arbor, MI : Cherry Lake Publishing, [2024] | Series: How can I help? | Audience: Grades 2-3 | Summary: "The World Wildlife Fund is dedicated to conserving nature and making the world safer for wildlife and wild places around the world. Readers will explore the history and impact of this organization and how it works with local communities and volunteers to make a global difference"– Provided by publisher.
Identifiers: LCCN 2024010025 | ISBN 9781668946824 (hardcover) | ISBN 9781668948217 (paperback) | ISBN 9781668949733 (ebook) | ISBN 9781668954294 (pdf)
Subjects: LCSH: World Wildlife Fund–Juvenile literature. | Nature conservation–Juvenile literature. | Wildlife conservation–Juvenile literature.
Classification: LCC QH75 .M3637 2024 | DDC 333.95/160601–dc23/eng/20240301
LC record available at https://lccn.loc.gov/2024010025

Cherry Lake Publishing Group would like to acknowledge the work of the Partnership for 21st Century Learning, a Network of Battelle for Kids. Please visit Battelle for Kids online for more information.

Printed in the United States of America

Note from publisher: Websites change regularly, and their future contents are outside of our control. Supervise children when conducting any recommended online searches for extended learning opportunities.

Cherry Lake Press

CONTENTS

Committed to Conservation 4

A History of Helping 8

Efforts to Save Earth 16

Glossary 22

Find Out More 23

Index 24

About the Author 24

COMMITTED TO CONSERVATION

To the people of China, the giant panda is a national treasure. Unfortunately, this peaceful bear is in danger. Hunting and **habitat** loss have reduced the giant panda population to less than 2,000 in the wild.

World Wildlife Fund (WWF) is determined to change this situation. WWF is the world's largest **conservation** organization.

WWF lets people know about their mission in different ways. This balloon is one of them!

Members of WWF carry out conservation efforts in 100 countries, including the United States and Canada. They work to restore, or bring back, animal populations. WWF also focuses on conserving a variety of freshwater, ocean, and forest habitats.

Look!

Take a closer look at the WWF logo. (A logo is a symbol used to identify an organization.) Why do you think WWF leaders chose to feature the giant panda?

Helping kids understand the importance of the environment is a big part of WWF's job.

Think!

WWF works to reduce people's ecological footprint, or impact on the environment. The size of an ecological footprint is based on humans' use of natural resources in a certain area. What activities do you think create a larger ecological footprint?

This organization aims to build a future that balances people's needs with respect for the environment and **biodiversity**.

The work WWF does makes a big difference. In the last 10 years, the giant panda has been moved off the **endangered** species list! But it is still **vulnerable**. Its population is growing, but it is still at risk. So are many other animals. WWF's work continues.

A HISTORY OF HELPING

In 1961, a handful of conservation groups existed across the globe. Yet most lacked a steady supply of money to fund their efforts.

Sixteen of the world's leading conservationists tried to solve this problem by forming WWF. It served as a **nonprofit** fund-raising organization. WWF's founders agreed that it would provide financial support to conservation groups around the globe.

Make a Guess!

Are you able to guess how many conservation projects WWF has invested in? Guess big! Since 1961, WWF has invested in over 10,000 projects in 157 countries.

WWF works with many other conservation groups, like the Wildlife Conservation Society at the Bronx Zoo.

At one time, there were only 14 red wolves left in the world. There are now around 300.

Ask Questions!

Want to find out what other species WWF helps protect? How about which specific habitats it works to conserve? With an adult, head online or contact a WWF office to learn more.

At first, WWF mainly funded projects to protect animals that were being threatened by human activities. The Tule goose in Canada and the red wolf in the United States were two examples of such species. By the early 1970s, however, WWF had broadened its focus. The organization started to set up national parks and nature reserves. This allowed WWF to conserve both animals and their natural habitats.

There are a lot of different ways to practice conservation and sustainable living in your town!

Today, WWF funds more funds over 3,000 active projects at any given time. In addition to partnering with other conservation groups, it **collaborates** with governments, businesses, and local communities. WWF efforts include conserving habitats. They also involve protecting wildlife, finding responsible ways to produce food, and studying how human activities affect Earth's climate.

Think!

WWF supports programs that promote sustainable living. Such programs encourage people to make choices that use up less of Earth's natural resources. Recycling is one activity that goes hand in hand with sustainable living. Are you able to think of any others?

WWF depends on both volunteers and paid employees. Some are scientists who conduct environmental research. Others are educators who create teaching materials that help students learn more about conservation.

Certain WWF workers are lobbyists. They try to influence government officials on issues that

Testing water and soil helps scientists learn more about an animal's habitat.

impact wildlife and natural habitats. Finally, WWF relies on financial experts and people who work in the media. They help fund the organization's projects and spread awareness.

Look!

With an adult, look for photographs of WWF workers in action! (Hint: The library and Internet are good places to start your search.) What are the people in those pictures doing? How are they supporting conservation?

EFFORTS TO SAVE EARTH

WWF uses several methods to support conservation. It builds partnerships with businesses to encourage sustainable living and to save various species and natural settings. Under WWF's guidance, companies learn to develop more environmentally-friendly production processes. In turn, these businesses sometimes donate to WWF projects across the globe.

Large businesses can make a big difference in conservation efforts, but individuals can help, too!

Make a Guess!

Are you able to guess how WWF pays for its conservation efforts? (Hint: There's more than one source!) Most of the organization's funding comes from individual contributions. Governments and businesses also provide financial support to WWF.

WWF thinks it is important to conserve the environment for the next generation.

Ask Questions!

What is WWF doing to save the Amazon rain forest? How is it encouraging cities around the world to conserve energy? Why is WWF concerned about marine turtles? With an adult, head online to learn the answers to these and other questions!

Members of WWF work closely with government officials, as well. They help shape laws that protect wildlife and the environment.

WWF is involved in a great deal of research. Conservationists study endangered animals and the condition of natural habitats. They track environmental challenges within communities. WWF helps residents use natural resources more carefully.

Helping to clean up your community helps WWF reach its goal.

Finally, WWF believes strongly in the value of public education. It hosts workshops, publishes a magazine, and sponsors television ads. WWF even sells art, clothes, and toys that are designed to teach the importance of conservation.

There's no guarantee that Earth's natural treasures will last forever. Fortunately, WWF is dedicated to helping people learn that lesson before it's too late.

Create!

Want to do your part to support WWF? Grab some old magazines, scissors, and glue. Create a collage of images that are related to WWF's conservation efforts. Try to find pictures of wildlife, natural habitats, and people practicing sustainable living!

GLOSSARY

biodiversity **(bye-oh-duh-VUHR-suh-tee)** the existence of many different kinds of plants and animals in a certain environment

collaborates **(kuh-LAH-buh-raytes)** works with another person or group in order to achieve something

conservation **(kahn-suhr-VAY-shuhn)** the act of protecting the environment and the plants and animals that live in it

endangered **(en-DAYN-jurd)** having a very high risk of extinction

habitat **(HAB-uh-tat)** an environment where certain plants and animals are usually found

natural resources **(NA-chuh-ruhl REE-sors-uhz)** materials or substances such as water and forests that occur in nature and are useful to humans

nonprofit **(nahn-PRAH-fit)** not existing for the main purpose of earning more money than is spent

reserves **(ri-ZURVZ)** areas of land or water that are set aside for conservation purposes

vulnerable **(VUHL-nuh-ruh-buhl)** having a high risk of extinction

FIND OUT MORE

Books

Hawley, Ella. *Exploring Our Impact on the Environment*. New York: PowerKids Press, 2013.

Littlewood, Peter. *Endangered Species*. Mankato, MN: Smart Apple Media, 2012.

Trueit, Trudi Strain. *Giant Pandas*. Mankato, MN: Amicus High Interest/ Amicus Ink, 2016.

Websites

With an adult, learn more online with these suggested searches.

National Geographic Kids—Giant Panda

Read more about the incredible animal featured on WWF's logo.

WWF—About Our Earth

Learn more about WWF, as well as Earth's many natural treasures!

INDEX

biodiversity, 7
climate, 13
conservation, 4–9, 13–21
ecological footprint, 7
endangered animals, 7, 19
environment, 6, 7, 16, 19
extinction, 4
food production, 13
habitat, 4, 6, 11, 13, 15, 19
hunting, 4
laws, 19
natural resources, 7, 13, 19
panda bear, 4–7
recycling, 13
research, 19
sustainable living, 13, 16, 21
threatened animals, 11
wildlife protection, 10–13
World Wildlife Fund history, 8–15
  how it pays for projects, 17
  how it supports conservation, 16–21
  how many projects it funds, 9, 13
  what it does, 4–7
  workers, 14

ABOUT THE AUTHOR

Katie Marsico is the author of more than 200 children's books. She lives in a suburb of Chicago, Illinois, with her husband and children.